GOD'S DESIGN® FOR LIFE

THE WORLD OF PLANTS

TEACHER SUPPLEMENT

1:1

▼**answersingenesis**

Petersburg, Kentucky, USA

ANSWERS IN GENESIS SCIENCE BY DEBBIE & RICHARD LAWRENCE

God's Design for Life
The World of Plants Teacher Supplement

© 2008 by Debbie & Richard Lawrence

Published by Answers in Genesis, 2800 Bullittsburg Church Rd., Petersburg KY 41080

You may contact the authors at (970) 686-5744.

ISBN: 1-60092-289-9

Cover design & layout: Diane King
Editors: Lori Jaworski, Gary Vaterlaus

Printed in China

www.answersingenesis.org   www.godsdesignscience.com

TABLE OF CONTENTS

WELCOME TO GOD'S DESIGN® FOR LIFE

God's Design for Life is a series that has been designed for use in teaching life science to elementary and middle school students. It is divided into three books: *The World of Plants, The World of Animals*, and *The Human Body*. Each book has 35 lessons including a final project that ties all of the lessons together.

In addition to the lessons, special features in each book include biographical information on interesting people as well as fun facts to make the subject more fun.

Although this is a complete curriculum, the information included here is just a beginning, so please feel free to add to each lesson as you see fit. A resource guide is included in the appendices to help you find additional information and resources. A list of supplies needed is included at the beginning of each lesson, while a master list of all supplies needed for the entire series can be found in the appendices.

Answer keys for all review questions, worksheets, quizzes, and the final exam are included here. Reproducible student worksheets and tests may be found on the supplementary CD-Rom for easy printing. Please contact Answers in Genesis if you wish to purchase a printed version of all the student materials, or go to www.AnswersBookstore.com.

If you wish to get through all three books of the *Life* series in one year, plan on covering approximately three lessons per week. The time required for each lesson varies depending on how much additional information you include, but plan on 20 minutes per lesson for beginners (grades 1–2) and 40 to 45 minutes for grades 3–8.

Quizzes may be given at the conclusion of each unit and the final exam may be given after lesson 34.

If you wish to cover the material in more depth, you may add additional information and take a longer period of time to cover all the material, or you could choose to do only one or two of the books in the series as a unit study.

WHY TEACH LIFE SCIENCE?

Maybe you hate science or you just hate teaching it. Maybe you love science but don't quite know how to teach it to your children. Maybe science just doesn't seem as important as some of those other subjects you need to teach. Maybe you need a little motivation. If any of these descriptions fits you, then please consider the following.

It is not uncommon to question the need to teach your kids hands-on science in elementary school. We could argue that the knowledge gained in science will be needed later in life in order for your children to be more productive and well-rounded adults. We could argue that teaching your children science also teaches them logical and inductive thinking and reasoning skills, which are tools they will need to be more successful. We could argue that science is a necessity in this technological world in which we live. While all of these arguments are true, not one of them is the real reason that we should teach our children science. The most important reason to teach science in elementary school is to give your children an understanding that God is our Creator, and the Bible can be trusted. Teaching science from a creation perspective is one of the best ways to reinforce your children's faith in God and to help them counter the evolutionary propaganda they face every day.

God is the Master Creator of everything. His handiwork is all around us. Our Great Creator put in place all of the laws of physics, biology, and chemistry. These laws were put here for us to see His wisdom and power. In science, we see the hand of God at work more than in any other subject. Romans 1:20 says, "For since the creation of the world His invisible attributes are clearly seen, being understood by the things that are made, even His eternal power and Godhead, so that they [men] are without excuse." We need to help our children see God as Creator of the world around them so they will be able to recog-nize God and follow Him.

The study of life science helps us understand the balance of nature so that we can be good stewards of our bodies, the plants, and the animals around us. It helps us appreciate the intricacies of life and the wonders of God's creation. Understanding the world of living things from a biblical point of view will prepare our children to deal with an ecology-obsessed world. It is critical to teach our children the truth of the Bible, how to evaluate the evidence, how to distinguish fact from theory and to realize that the evidence, rightly interpreted, supports biblical creation, not evolution.

It's fun to teach life science! It's interesting, too. Children have a natural curiosity about living things, so you won't have to coax them to explore the world of living creatures. You just have to direct their curiosity and reveal to them how interesting life science can be.

Finally, teaching life science is easy. It's all around us. Everywhere we go, we are surrounded by living things. You won't have to try to find strange materials for experiments or do dangerous things to learn about life.

HOW DO I TEACH SCIENCE?

In order to teach any subject, you need to understand that people learn in different ways. Most people, and children in particular, have a dominant or preferred learning style in which they absorb and retain information more easily.

If a student's dominant style is:

Auditory
He needs not only to hear the information but he needs to hear himself say it. This child needs oral presentation as well as oral drill and repetition.

Visual
She needs things she can see. This child responds well to flashcards, pictures, charts, models, etc.

Kinesthetic
He needs active participation. This child remembers best through games, hands-on activities, experiments, and field trips.

Also, some people are more relational while others are more analytical. The relational student needs to know why this subject is important and how it will affect him personally. The analytical student, however, wants just the facts.

If you are trying to teach more than one student, you will probably have to deal with more than one learning style. Therefore, you need to present your lessons in several different ways so that each student can grasp and retain the information.

GRADES 1–2

Because *God's Design Science* books are designed to be used with students in grades 1–8, each lesson has been divided into three sections. The "Beginner" section is for students in grades 1–2. This part contains a read-aloud section explaining the material for that lesson followed by a few questions to make sure that the students understand what they just heard. We recommend that you do the hands-on activity in the blue box in the main part of the lesson to help your students see and understand the concepts.

GRADES 3–8

The second part of each lesson should be completed by all upper elementary and junior high students. This is the main part of the lesson containing a reading section, a hands-on activity that reinforces the ideas in the reading section (blue box), and a review section that provides review questions and application questions (red box).

GRADES 6–8

Finally, for middle school/junior high age students, we provide a "Challenge" section that contains more challenging material as well as additional activities and projects for older students (green box).

We have included periodic biographies to help your students appreciate the great men and women who have gone before us in the field of science.

We suggest a threefold approach to each lesson:

Introduce the topic

We give a brief description of the facts. Frequently you will want to add more information than the essentials given in this book. In addition to reading this section aloud, you may wish to do one or more of the following:

- Read a related book with your students.
- Write things down to help your visual students.
- Give some history of the subject. We provide some historical sketches to help you, but you may want to add more.
- Ask questions to get your students thinking about the subject.
- The "FUN FACT" section adds fun or interesting information.

Make observations and do experiments

- Hands-on projects are suggested for each lesson. This part of each lesson may require help from the teacher.
- Have your students perform the activity by themselves whenever possible.

Review

- The "What did we learn?" section has review questions.
- The "Taking it further" section encourages students to
  - Draw conclusions
  - Make applications of what was learned
  - Add extended information to what was covered in the lesson

---

By teaching all three parts of the lesson, you will be presenting the material in a way that children with any learning style can both relate to and remember.

Also, this approach relates directly to the scientific method and will help your students think more scientifically. The *scientific method* is just a way to examine a subject logically and learn from it. Briefly, the steps of the scientific method are:

1. Learn about a topic.
2. Ask a question.
3. Make a hypothesis (a good guess).
4. Design an experiment to test your hypothesis.
5. Observe the experiment and collect data.
6. Draw conclusions. (Does the data support your hypothesis?)

Note: It's okay to have a "wrong hypothesis." That's how we learn. Be sure to help your students understand why they sometimes get a different result than expected.

Our lessons will help your students begin to approach problems in a logical, scientific way.

HOW DO I TEACH CREATION VS. EVOLUTION?

We are constantly bombarded by evolutionary ideas about living things in books, movies, museums, and even commercials. These raise many questions: Did dinosaurs really live millions of years ago? Did man evolve from apes? Which came first, Adam and Eve or the cavemen? Where did living things come from in the first place? The Bible answers these questions and this book accepts the historical accuracy of the Bible as written. We believe this is the only way we can teach our children to trust that everything God says is true.

There are five common views of the origins of life and the age of the earth:

Historical biblical account	Progressive creation	Gap theory	Theistic evolution	Naturalistic evolution
Each day of creation in Genesis is a normal day of about 24 hours in length, in which God created everything that exists. The earth is only thousands of years old, as determined by the genealogies in the Bible.	The idea that God created various creatures to replace other creatures that died out over millions of years. Each of the days in Genesis represents a long period of time (day-age view) and the earth is billions of years old.	The idea that there was a long, long time between what happened in Genesis 1:1 and what happened in Genesis 1:2. During this time, the "fossil record" was supposed to have formed, and millions of years of earth history supposedly passed.	The idea that God used the process of evolution over millions of years (involving struggle and death) to bring about what we see today.	The view that there is no God and evolution of all life forms happened by purely naturalistic processes over billions of years. Ken Ham et al., *The Answers Book*, (El Cajon: Master Books, 2000), 33–76.

Any theory that tries to combine the evolutionary time frame with creation presupposes that death entered the world before Adam sinned, which contradicts what God has said in His Word. The view that the earth (and its "fossil record") is hundreds of millions of years old damages the gospel message. God's completed creation was "very good" at the end of the sixth day (Genesis 1:31). Death entered this perfect paradise *after* Adam disobeyed God's command. It was the punishment for Adam's sin (Genesis 2:16–17; 3:19; Romans 5:12–19). Thorns appeared when God cursed the ground because of Adam's sin (Genesis 3:18).

The first animal death occurred when God killed at least one animal, shedding its blood, to make clothes for Adam and Eve (Genesis 3:21). If the earth's "fossil record" (filled with death, disease, and thorns) formed over millions of years before Adam appeared (and before he sinned), then death no longer would be the penalty for sin. Death, the "last enemy" (1 Corinthians 15:26), diseases (such as cancer), and thorns would instead be part of the original creation that God labeled "very good." No, it is clear that the "fossil record" formed some time *after* Adam sinned—not many millions of years before. Most fossils were formed as a result of the worldwide Genesis Flood.

When viewed from a biblical perspective, the scientific evidence clearly supports a recent creation by God, and not naturalistic evolution and millions of years. The volume of evidence supporting the biblical creation account is substantial and cannot be adequately covered in this book. If you would like more information on this topic, please see the resource guide in the appendices. To help get you started, just a few examples of evidence supporting biblical creation are given below:

Evolutionary Myth: Humans have been around for more than one million years.

The Truth: If people have been on earth for a million years, there would be trillions of people on the earth today, even if we allowed for worst-case plagues, natural disasters, etc. The number of people on earth today is about 6.5 billion. If the population had grown at only a 0.01% rate (today's rate is over 1%) over 1 million years, there could be 10^{43} people today (that's a number with 43 zeros after it)! Repopulating the earth after the Flood would only require a population growth rate of 0.5%, half of what it is today.

John D. Morris, *The Young Earth* (Colorado Springs: Creation Life Publishers, 1994), 70–71. See also "Billions of People in Thousands of Years?" at www.answersingenesis.org/go/billions-of-people.

Evolutionary Myth: Man evolved from an ape-like creature.

The Truth: All so-called "missing links" showing human evolution from apes have been shown to be either apes, humans, or deliberate hoaxes. These links remain missing.

Duane T. Gish, *The Amazing Story of Creation from Science and the Bible* (El Cajon: Institute for Creation Research, 1990), 78–83.

Evolutionary Myth: All animals evolved from lower life forms.

The Truth: While Darwin predicted that the fossil record would show numerous transitional fossils, even more than 145 years later, all we have are a handful of disputable examples. For example, there are no fossils showing something that is part way between a dinosaur and a bird. Fossils show that a snail has always been a snail; a squid has always been a squid. God created each animal to reproduce after its kind (Genesis 1:20–25).

Ibid., p. 36, 53–60.

> **Evolutionary Myth:** Dinosaurs evolved into birds.

> **The Truth:** Flying birds have streamlined bodies, with the weight centralized for balance in flight; hollow bones for lightness, which are also part of their breathing system; powerful muscles for flight; and very sharp vision. And birds have two of the most brilliantly-designed structures in nature—their feathers and special lungs. It is impossible to believe that a reptile could make that many changes over time and still survive.
>
> Gregory Parker et al., *Biology: God's Living Creation* (Pensacola: A Beka Books, 1997) 474–475.

> **Evolutionary Myth:** Thousands of changes over millions of years resulted in the creatures we see today.

> **The Truth:** What is now known about human and animal anatomy shows the body structures, from the cells to systems, to be infinitely more complex than was believed when Darwin published his work in 1859. Many biologists and especially microbiologists are now saying that there is no way these complex structures could have developed by natural processes.
>
> Ibid., p. 384–385.

Since the evidence does not support their theories, evolutionists are constantly coming up with new ways to try to support what they believe. One of their ideas is called punctuated equilibrium. This theory of evolution says that rapid evolution occurred in small isolated populations, and left no evidence in the fossil record. There is no evidence for this, nor any known mechanism to cause these rapid changes. Rather, it is merely wishful thinking. We need to teach our children the difference between science and wishful thinking.

Despite the claims of many scientists, if you examine the evidence objectively, it is obvious that evolution and millions of years have not been proven. You can be confident that if you teach that what the Bible says is true, you won't go wrong. Instill in your student a confidence in the truth of the Bible in all areas. If scientific thought seems to contradict the Bible, realize that scientists often make mistakes, but God does not lie. At one time scientists believed that the earth was the center of the universe, that living things could spring from non-living things, and that blood-letting was good for the body. All of these were believed to be scientific facts but have since been disproved, but the Word of God remains true. If we use modern "science" to interpret the Bible, what will happen to our faith in God's Word when scientists change their theories yet again?

INTEGRATING THE SEVEN C'S

Throughout the *God's Design® for Science* series you will see icons that represent the Seven C's of History. The Seven C's is a framework in which all of history, and the future to come, can be placed. As we go through our daily routines we may not understand how the details of life connect with the truth that we find in the Bible. This is also the case for students. When discussing the importance of the Bible you may find yourself telling students that the Bible is relevant in everyday activities. But how do we help the younger generation see that? The Seven C's are intended to help.

The Seven C's can be used to develop a biblical worldview in students, young or old. Much more than entertaining stories and religious teachings, the Bible has real connections to our everyday life. It may be hard, at first, to see how many connections there are, but with practice ,the daily relevance of God's Word will come alive. Let's look at the Seven C's of History and how each can be connected to what the students are learning.

CREATION ✓

God perfectly created the heavens, the earth, and all that is in them in six normal-length days around 6,000 years ago.

This teaching is foundational to a biblical worldview and can be put into the context of any subject. In science, the amazing design that we see in nature—whether in the veins of a leaf or the complexity of your hand—is all the handiwork of God. Virtually all of the lessons in *God's Design for Science* can be related to God's creation of the heavens and earth.

Other contexts include:

Natural laws—any discussion of a law of nature naturally leads to God's creative power.

DNA and information—the information in every living thing was created by God's supreme intelligence.

Mathematics—the laws of mathematics reflect the order of the Creator.

Biological diversity—the distinct kinds of animals that we see were created during the Creation Week, not as products of evolution.

Art—the creativity of man is demonstrated through various art forms.

History—all time scales can be compared to the biblical time scale extending back about 6,000 years.

Ecology—God has called mankind to act as stewards over His creation.

CORRUPTION ✓

After God completed His perfect creation, Adam disobeyed God by eating the forbidden fruit. As a result, sin and death entered the world, and the world has been in decay since that time. This point is evident throughout the world that we live in. The struggle for survival in animals, the death of loved ones, and the violence all around us are all examples of the corrupting influence of sin.

Other contexts include:

Genetics—the mutations that lead to diseases, cancer, and variation within populations are the result of corruption.

Biological relationships—predators and parasites result from corruption.

History—wars and struggles between mankind, exemplified in the account of Cain and Abel, are a result of sin.

CATASTROPHE

God was grieved by the wickedness of mankind and judged this wickedness with a global Flood. The Flood covered the entire surface of the earth and killed all air-breathing creatures that were not aboard the Ark. The eight people and the animals aboard the Ark replenished the earth after God delivered them from the catastrophe.

The catastrophe described in the Bible would naturally leave behind much evidence. The studies of geology and of the biological diversity of animals on the planet are two of the most obvious applications of this event. Much of scientific understanding is based on how a scientist views the events of the Genesis Flood.

Other contexts include:

Biological diversity—all of the birds, mammals, and other air-breathing animals have populated the earth from the original kinds which left the Ark.

Geology—the layers of sedimentary rock seen in roadcuts, canyons, and other geologic features are testaments to the global Flood.

Geography—features like mountains, valleys, and plains were formed as the floodwaters receded.

Physics—rainbows are a perennial sign of God's faithfulness and His pledge to never flood the entire earth again.

Fossils—Most fossils are a result of the Flood rapidly burying plants and animals.

Plate tectonics—the rapid movement of the earth's plates likely accompanied the Flood.

Global warming/Ice Age—both of these items are likely a result of the activity of the Flood. The warming we are experiencing today has been present since the peak of the Ice Age (with variations over time).

Confusion

God commanded Noah and his descendants to spread across the earth. The refusal to obey this command and the building of the tower at Babel caused God to judge this sin. The common language of the people was confused and they spread across the globe as groups with a common language. All people are truly of "one blood" as descendants of Noah and, originally, Adam.

The confusion of the languages led people to scatter across the globe. As people settled in new areas, the traits they carried with them became concentrated in those populations. Traits like dark skin were beneficial in the tropics while other traits benefited populations in northern climates, and distinct people groups, not races, developed.

Other contexts include:

Genetics—the study of human DNA has shown that there is little difference in the genetic makeup of the so-called "races."

Languages—there are about seventy language groups from which all modern languages have developed.

Archaeology—the presence of common building structures, like pyramids, around the world confirms the biblical account.

Literature—recorded and oral records tell of similar events relating to the Flood and the dispersion at Babel.

Christ

God did not leave mankind without a way to be redeemed from its sinful state. The Law was given to Moses to show how far away man is from God's standard of perfection. Rather than the sacrifices, which only covered sins, people needed a Savior to take away their sin. This was accomplished when Jesus Christ came to earth to live a perfect life and, by that obedience, was able to be the sacrifice to satisfy God's wrath for all who believe.

The deity of Christ and the amazing plan that was set forth before the foundation of the earth is the core of Christian doctrine. The earthly life of Jesus was the fulfillment of many prophecies and confirms the truthfulness of the Bible. His miracles and presence in human form demonstrate that God is both intimately concerned with His creation and able to control it in an absolute way.

Other contexts include:

Psychology—popular secular psychology teaches of the inherent goodness of man, but Christ has lived the only perfect life. Mankind needs a Savior to redeem it from its unrighteousness.

Biology—Christ's virgin birth demonstrates God's sovereignty over nature.

Physics—turning the water into wine and the feeding of the five thousand demonstrate Christ's deity and His sovereignty over nature.

History—time is marked (in the western world) based on the birth of Christ despite current efforts to change the meaning.

Art—much art is based on the life of Christ and many of the masters are known for these depictions, whether on canvas or in music.

CROSS

Because God is perfectly just and holy, He must punish sin. The sinless life of Jesus Christ was offered as a substitutionary sacrifice for all of those who will repent and put their faith in the Savior. After His death on the Cross, He defeated death by rising on the third day and is now seated at the right hand of God.

The events surrounding the crucifixion and resurrection have a most significant place in the life of Christians. Though there is no way to scientifically prove the resurrection, there is likewise no way to prove the stories of evolutionary history. These are matters of faith founded in the truth of God's Word and His character. The eyewitness testimony of over 500 people and the written Word of God provide the basis for our belief.

Other contexts include:

Biology—the biological details of the crucifixion can be studied alongside the anatomy of the human body.

History—the use of crucifixion as a method of punishment was short-lived in historical terms and not known at the time it was prophesied.

Art—the crucifixion and resurrection have inspired many wonderful works of art.

CONSUMMATION

God, in His great mercy, has promised that He will restore the earth to its original state—a world without death, suffering, war, and disease. The corruption introduced by Adam's sin will be removed. Those who have repented and put their trust in the completed work of Christ on the Cross will experience life in this new heaven and earth. We will be able to enjoy and worship God forever in a perfect place.

This future event is a little more difficult to connect with academic subjects. However, the hope of a life in God's presence and in the absence of sin can be inserted in discussions of human conflict, disease, suffering, and sin in general.

Other contexts include:

History—in discussions of war or human conflict the coming age offers hope.

Biology—the violent struggle for life seen in the predator-prey relationships will no longer taint the earth.

Medicine—while we struggle to find cures for diseases and alleviate the suffering of those enduring the effects of the Curse, we ultimately place our hope in the healing that will come in the eternal state.

The preceding examples are given to provide ideas for integrating the Seven C's of History into a broad range of curriculum activities. We would recommend that you give your students, and yourself, a better understanding of the Seven C's framework by using AiG's *Answers for Kids* curriculum. The first seven lessons of this curriculum cover the Seven C's and will establish a solid understanding of the true history, and future, of the universe. Full lesson plans, activities, and student resources are provided in the curriculum set.

We also offer bookmarks displaying the Seven C's and a wall chart. These can be used as visual cues for the students to help them recall the information and integrate new learning into its proper place in a biblical worldview.

Even if you use other curricula, you can still incorporate the Seven C's teaching into those. Using this approach will help students make firm connections between biblical events and every aspect of the world around them, and they will begin to develop a truly biblical worldview and not just add pieces of the Bible to what they learn in "the real world."

UNIT 1
INTRODUCTION TO LIFE SCIENCE

LESSON 1

IS IT ALIVE?

BIOLOGY IS THE STUDY OF LIVING THINGS.

SUPPLY LIST

Copy of "Is It Alive?" worksheet
Six items to display/discuss: some living, some non-living (book, pet, can, eraser, plant, etc.)

BEGINNERS

- If something is alive, what are some things that it will do? **Eat, breathe, move, grow, reproduce, etc.**
- What are the building blocks for plants and animals? **Cells.**
- Is it alive? **Answers will vary.**

WHAT DID WE LEARN?

- What are the six questions you should ask to determine if something is alive? **Does it eat?, Does it breathe?, Does it grow?, Does it reproduce?, Can it move?, Does it have cells?**

TAKING IT FURTHER

- Is a piece of wood that has been cut off of a tree living? (Hint: Is it growing? Can it respond?) **No, it is not living anymore; although the tree it came from may still be living.**
- Is paper alive? **No. It is made from wood but it is not alive.**
- Is a seed alive? **This is a harder question. A seed has the potential for life, but it is not growing. You have to decide for yourself.**

LESSON 2

WHAT IS A KINGDOM?

IT'S ALIVE, BUT WHAT IS IT?

SUPPLY LIST

Poster board   Copy of "Clue Cards"   Pen   Scissors

BEGINNERS

- Which living things can make their own food—plants or animals? **Plants.**
- Why do plants need sunshine? **They use sunlight to make food.**
- Name one important difference between plants and animals. **Accept reasonable answers.**

CLUE CARDS

- **Plants only—Chlorophyll, photosynthesis, needs sun, cannot move around, needs carbon dioxide, created on the 3rd day of creation. Animals only—Moves around, cannot make food, carbon dioxide is a waste product, no chlorophyll. Both—Alive, cells, reproduces same kind, needs oxygen, designed by God, eaten by animals.**

WHAT DID WE LEARN?

- What do plants and animals have in common? **God created them all, all are alive, all have cells, all reproduce their own kind, and all need oxygen.**
- What makes plants unique? **They have chlorophyll, perform photosynthesis, and cannot move freely.**
- What makes animals unique? **They cannot produce their own food and can move freely.**

TAKING IT FURTHER

- Are mushrooms plants? **No, they do not have chlorophyll or perform photosynthesis.**
- Why do you think they are or are not? **Fungi have most of the characteristics of plants, but do not have chlorophyll and can live without sunlight. This is why scientists now group them in their own kingdom.**

LESSON
3

CLASSIFICATION SYSTEM

TAXONOMY—CLASSIFICATION OF LIVING THINGS

SUPPLY LIST

Plant and animal guides or an encyclopedia

BEGINNERS

- How do scientists split living things into groups? **They examine what is the same and what is different and put things that are similar in the same group.**
- What is the first major division of living things called? **A kingdom.**
- What are the two main groups of animals? **Those with backbones and those without backbones.**

WHAT DID WE LEARN?

What are the five kingdoms recognized today? **Plants, animals, fungi, protists, and monerans.**
- How do scientists determine how to classify a living thing? **They look at common characteristics and at different characteristics.**
- What are the seven levels of the classification system? **Kingdom, phylum, class, order, family, genus, and species.**

TAKING IT FURTHER

- Why can pet dogs breed with wild wolves? **They are both the same kind of animal. Wolves, jackals, coyotes, wild dogs, and domestic dogs all came from the same ancestors. If any two animals can produce fertile offspring then they are most likely from the same animal kind. Wolves don't generally breed with domestic dogs because of their location and habits, but biologically they are the same kind of animal.**

- How many of each animal did Noah take on the Ark? **Two of some animals and seven of other animals (see Genesis 7). Noah would only have taken two canines (dogs) on the Ark. Afterwards, the offspring of those two dogs resulted in the wide variety of dogs we see today.**

CHALLENGE: PLANT CLASSIFICATION

Note: The third paragraph of the Challenge section should be replaced with this one.

Nonvascular plants are divided into three groups: mosses, liverworts, and hornworts. Together these are called bryophytes. These nonvascular plants have leaves and stems, but do not have true roots. They reproduce by spores, not with flowers. The bryophytes tend to grow in clumps in moist areas. You may find them growing on tree trunks or along streams, but don't confuse them with the algae growing in the water. Even though algae contain chlorophyll, they are not plants since they do not have leaves, stems, and roots.

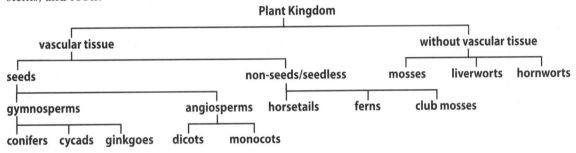

PLANT & ANIMAL CELLS

LESSON 4

THE SMALLEST UNIT OF LIFE

SUPPLY LIST

Option A: Paper models

    Colored construction paper   Scissors   Glue

Option B: 3-D models (messier but more fun)

    For each child: Small shoe box   1 qt. plastic zipper bag   Several green grapes   Several raisins 1 large red grape or marble   For everyone to use: Yellow gelatin mix (Note: mix this gelatin according to the package directions about an hour before you plan to do the project.)

Supplies for Challeng: Microscope   Slides   Onion   Sharp knife

BEGINNERS

- What shape are most animal cells? **Round.**
- What shape are most plant cells? **Rectangular.**

- What is the job of the cell membrane? **It is like skin and holds the cell together.**
- What is the job of the nucleus? **It is the brain—it tells everything else what to do.**
- What can plant cells do that animal cells cannot do? **They turn sunlight into food.**

WHAT DID WE LEARN?

- What parts or structures do all plant and animal cells have? **Cell membrane, nucleus, vacuoles, mitochondria, and cytoplasm.**
- What structures are unique to plants? **Cell wall and chloroplasts.**
- What distinguishes animal cells from plant cells? Possible answers: **Plant cells can perform photosynthesis and have cell walls but animal cells do not; their shape is different.**

TAKING IT FURTHER

- A euglena is a single-celled living organism that can move around by itself. It eats other creatures, but it also has chlorophyll in its cell. Is it a plant, an animal, or something else? **Scientists do not agree on this and other unusual creatures. They usually put them in their own category, called protists.**

QUIZ 1
INTRODUCTION TO LIFE SCIENCE
LESSONS 1–4

Mark each statement as either True or False.

1. _T_ All living creatures have cells.
2. _F_ Plants do not need oxygen.
3. _T_ Growth and change can be signs of life.
4. _F_ Non-living things absorb nutrients.
5. _F_ Plants cannot move so they are not alive.
6. _T_ A kingdom is a way to group things together by similar characteristics.
7. _F_ Plants and protists are the two main kingdoms of living things.
8. _F_ Plants and animals both have chlorophyll.
9. _T_ Vacuoles store food inside of cells.
10. _T_ The nucleus is the control center of a cell.

Short answer:

11. Name three differences between plant and animal cells. **Shape of the cells—plant cells are usually rectangular, animal cells are usually round; plant cells have chlorophyll, animal cells do not; plant cells have a cell wall, animal cells do not; only plant cells perform photosynthesis.**

12. Describe how to tell if something is alive. **Eats, breathes, grows, reproduces, moves/responds to its environment, and has cells.**

CHALLENGE QUESTIONS

Fill in the blanks using the terms below. Not all words are used.

13. The Law of **_biogenesis_** states that life always comes from life.

14. _**Chemical evolution**_ says that life originally came from non-living chemicals.

15. During _**mitosis**_ a cell divides into two identical cells.

16. A _**conifer or gymnosperm**_ is a type of plant that produces seeds in cones.

17. A _**Ginkgo**_ tree is sometimes called a living fossil.

18. _**Angiosperms**_ produce seeds that are enclosed in fruit.

19. The belief that life springs up from its environment is called _**spontaneous generation**_.

20. During _**metaphase**_ the chromosomes in a cell line up in the middle.

21. During _**anaphase**_ the duplicate chromosomes are pulled apart.

22. _**Cytokinesis**_ occurs when the cytoplasm in a cell is divided.

23. A _**dicot**_ has a seed with two parts.

24. A _**monocot**_ has a seed with only one part.

25. A cycad is a type of _**gymnosperm**_.

Unit 2
Flowering Plants & Seeds

LESSON 5

Flowering Plants

God's gift of life to the world

Supply list

A field guide for flowers   Access to several flowering plants (What you have growing in your garden or yard is probably sufficient.)

Supplies for Challenge: Research materials—depends on your selected topic

Beginners

- What are the four parts of plants? **Roots, stems, leaves, and flowers.**
- What is the job of each of the parts? **Roots hold the plant in place and suck up water and nutrients. Stems help the plant stand up and to move water and nutrients inside the plant. Leaves turn sunlight into food. Flowers produce seeds.**

What did we learn?

- What are the four major parts of a plant? **Roots, stem, leaves and flowers.**
- What is the purpose for each part? **Roots hold the plant in place and suck up water and nutrients. Stems help the plant stand up and to move water and nutrients inside the plant. Leaves turn sunlight into food. Flowers produce seeds.**

Taking it further

- What characteristics other than the flowers can be used to help identify a plant? **Leaves, fruit, and bark can all be used to identify a plant that is not in bloom.**
- What similarities did you notice between the flowers you examined? **Answers will vary.**
- What differences did you see? **Answers will vary.**
- Can you use size to determine what a plant is? Why or why not? (Hint: Is a tiny seedling just as much an oak tree as the giant oak that is 100 years old?) **Size alone cannot tell you what a plant is.**
- Why might you need to identify a plant? Possible answers: **To recognize poisonous plants such as poison ivy, to choose good plants for a garden or landscape, and to enjoy God's creation are just a few reasons.**

LESSON 6

GRASSES

DO I HAVE TO CUT IT AGAIN, MOM?

SUPPLY LIST

Magnifying glass   Grass plant

Supplies for Challenge: Kentucky bluegrass seeds   Corn seeds   Other grass seeds as available (wheat, oats, rye, fescue, etc.)   Baking dish   Potting soil   Craft sticks   Marker
Copy of "Grass Comparison" worksheet

BEGINNERS

- Why are different types of grass so important? **They provide food for people and animals.**
- Name one kind of grass eaten by animals. **Prairie grass, hay, timothy, corn.**
- Name one kind of grass eaten by people. **Wheat, oats, rice, and corn.**

WHAT DID WE LEARN?

- Name four types of grass. **Turf, cereal, forage, and ornamental.**
- Describe the roots of a grass plant. **Fibrous root system with many small roots going out in several directions.**
- Why are grasses so important? **They are a major food source for animals and humans.**

TAKING IT FURTHER

- Why can grass be cut over and over and still grow, while a tree that is cut down will die? **Recall that the leaves of the grass grow from the base of the plant. So cutting off the top of the leaves does not damage the growing center of the plant. However, trees grow at the ends of the stems and branches.**
- Why is grass so hard to get rid of in a flower garden? **Consider the root structure; its fibrous design helps the plant spread and survive.**
- What part of grass plants do humans eat? **They generally eat the seeds.**
- What part of grass plants do most animals eat? **They usually eat the leaves and the seeds.**
- Why can a cow eat certain grasses that you can't? **Cows have a very different digestive system that can break down the grass that humans can't digest.**

LESSON 7

TREES

DID GEORGE WASHINGTON REALLY CHOP DOWN THE CHERRY TREE?

SUPPLY LIST

Index cards labeled with vocabulary words   Markers or crayons

Supplies for Challenge: Drawing materials

BEGINNERS

- What are the two different groups of trees? **Deciduous and evergreen.**
- What kind of leaves do deciduous trees usually have? **Flat leaves.**
- What kind of leaves do evergreen trees usually have? **Needles.**
- What covers a tree trunk to help protect it? **Bark.**
- How can you tell how old a tree is? **By counting its growth rings.**

WHAT KIND OF TREE IS THIS?

- **Angiosperm, broad leaf, flowers, oak, maple, cherry should all have deciduous picture only. Gymnosperm, needles, cones, fir, pine, spruce, conifer should all have evergreen only. Seeds, bark growth rings should have both pictures. A few evergreen trees have broad leaves so both pictures could be acceptable for this word. A few trees that bear cones also lose their leaves, so coniferous could also have both pictures.**

WHAT DID WE LEARN?

- What makes a plant a tree? **A single woody stem, grows to be tall, needs no support, and has bark.**
- How are deciduous and evergreen trees different? **Deciduous trees lose their leaves in winter and grow new ones in spring. Evergreen trees do not lose their leaves in the winter.**
- How are angiosperms and gymnosperms different? **Angiosperms produce seeds with fruits and flowers. The seeds are located in the fruit. Gymnosperms produce seeds in cones.**

TAKING IT FURTHER

- Do evergreen trees have growth rings? **Even though evergreens do not lose their leaves, they take a break from growing in the winter so they do have growth rings.**
- How long do you think a tree lives? **Some trees only live a few years and some trees live to be hundreds, or even thousands, of years old. It depends on the variety of the tree and the growing conditions.**

LESSON 8

SEEDS

GERMINATION—THE BEGINNING OF LIFE

SUPPLY LIST

Copy of "Germination Data Sheet"   5 jars (1 with a lid)   Paper towels   Steel wool
15–20 dried bean seeds   Black construction paper   Tape

BEGINNERS

- What part of a plant grows into a new plant? **Seeds.**
- What three things must be present before a seed will start growing? **Water, warmth, and air/oxygen.**

WHAT DID WE LEARN?

- What conditions must be present for most seeds to sprout or germinate? **Water, oxygen, and warmth.**
- Is soil necessary for seeds to germinate? **No, you germinated them in wet paper towels.**

Taking it further

- If plants don't need soil to germinate, why do plants need soil to grow? **The seed has some stored energy that helps it get started. Once this energy is used up, the plant's roots must absorb nutrients from the soil, and the leaves will make food from the sun.**

- Our seeds germinated in the dark. Can the plants continue to grow in the dark? **No, once the seed's energy is used up, the plant needs sunlight to make more food.**

- Why do seeds require these three conditions to begin growing? **God designed them that way so that the seeds will wait until it is likely the plants will survive before they germinate. If seeds germinated in the cold, the plant may not survive because the roots would freeze. If seeds germinated without oxygen, the plant would not be able to grow and it would die. If seeds germinated without water, the plants would soon wither. If seeds continued to germinate in unfavorable conditions, many plants could become extinct.**

- How long can seeds remain dormant? **Some seeds have sprouted after 100 years or more of waiting for the right conditions.**

LESSON
9

Monocots & Dicots

What's inside that seed?

Supply list

Several bean seeds (pinto, kidney, etc.) *** Soak these seeds in water for approx. 24 hours prior to use.
Several corn seeds (not popcorn) *** Soak these seeds in water for approx. 24 hours prior to use.
Dissecting scalpel or very sharp knife (for adult use only)   Magnifying glass   Jar or plastic cup
Paper towels
Supplies for Challenge: Other seeds for dissection

Beginners

- Why can a seed grow for a little while when it is not planted in the soil? **The seed contains food that helps the new plant grow for a while.**

- What is on the outside of a seed? **The seed covering or seed coat that protects it.**

- How many pieces does a bean seed have? **Two.**

- How many pieces does a corn seed have? **One.**

Seed dissection

- Why does a seed need water to germinate? **The seed coat protects the seed until water is available. Then the water helps to soften the seed coat and begin the growing process.**

- How does the seed coat of the corn differ from the seed coat of the beans? **The bean seed coat should be removed easily, and split into two parts. The corn seed coat cannot be removed easily, and will not divide.**

What did we learn?

- What differences did you observe between the monocot and dicot seeds? **Monocots have one part and a tougher seed coat. Dicots have two parts and a softer seed coat.**

- What parts of each seed were you able to identify? **You should have been able to identify the seed coat, hilum, plumule, radicle, cotyledons and the endosperm of the corn seed.**

- What is the plumule? **Part that grows into the stem and leaves of the plant.**
- What is the radicle? **Part that grows into the roots of the plant.**
- What is the purpose of the cotyledon? **It provides energy for the new plant until the roots get big enough to begin supporting the plant, and sometimes becomes the first embryonic leaves of a seedling.**

TAKING IT FURTHER

- Why did you need to soak the seeds before dissecting them? **Moisture is needed to soften the seed coat and begin the germination process. This is one of God's ways of preserving the seed until conditions are good for the plant to grow. If seeds sprouted when there was no water available, the young plants would soon wither and die.**
- What differences do you think you might find in plants that grow from monocot and dicot seeds? **There are differences in the root structures, stems, leaves, and flowers. We will discuss some of these in later lessons.**

CHALLENGE: GERMINATION

- **Beans—epigeal germination, Corn—hypogeal germination.**

LESSON 10 SEEDS—WHERE ARE THEY?

HOW DO THEY GET AROUND?

SUPPLY LIST

Several different fruits and vegetables (apple, tomato, peach, etc.)   Baking dish lined with aluminum foil
Several pinecones with scales tightly shut   Copy of "Seeds Get Around" worksheet
Supplies for Challenge: Several different kinds of seeds   Whole coconut or coconut seed, if available
Copy of "Water Dispersal Test"

BEGINNERS

- Name three places you can find seeds. **Fruit, vegetables, pinecones, and dead flowers.**
- Name three ways that seeds get from the plant to another location. **Carried by an animal, blown by the wind, and exploding from the seed pod.**

SEEDS GET AROUND WORKSHEET

- **Ideas might include: Wind: dandelion—parachute, maple—helicopter; Animal: cocklebur—Velcro; Explosion: touch-me-not, violets, witch hazel—bomb.**

WHAT DID WE LEARN?

- What are three ways seeds can be moved or dispersed? **By the wind, by animals or by explosion.**
- Where are good places to look for seeds? **In fruit, pinecones, or "dead" flowers.**

TAKING IT FURTHER

- How do people aid in the dispersal of seeds? Possible answers: **Farming, gardening, hiking (on our clothes), etc.**

- What has man done to change or improve seeds or plants? Possible answers: **Man has used cross-pollination and genetic alteration of seeds to develop plants that are more resistant to disease or insects, that are higher in nutritional content, have larger blossoms or have different colors, as well as many other changes.**

- If a seed is small, will the mature plant also be small? **Not usually; plants often grow large, even if they start from a small seed.**

- Do the largest plants always have the largest seeds? **No.**

- Why do you think God created many large plants to have small seeds? **Small seeds are more easily dispersed.**

- Can you name a plant that disperses its seeds by the whole plant blowing around? **Tumbleweed.**

CHALLENGE: WATER DISPERSAL

- **Plants that are close together compete for room, nutrients, water, and sunlight. Seed dispersal allows plants to spread out. This gives the seeds a better chance to land in areas where there is more room, nutrients, water, and sunlight, thus allowing new plants a better chance of survival. The coconut can float because it is less dense than water, just like a ship.**

QUIZ 2 FLOWERING PLANTS & SEEDS

LESSONS 5–10

Match the term with its definition.

1. _A_ Monocot
2. _D_ Dicot
3. _B_ Cotyledon
4. _F_ Hilum
5. _I_ Plumule
6. _G_ Radicle
7. _C_ Seed Coat
8. _E_ Deciduous
9. _H_ Evergreen

Answer yes or no. Would a seed geminate if placed in each of the following conditions? If no, explain what is missing.

10. A seed planted in a garden in the spring time and watered every day. **Yes—All necessary conditions are present.**

11. A seed in a desert. **Probably not—There is probably not enough water.**

12. A seed planted in the dirt on the moon. **No—There is not enough oxygen, moisture, or heat on the moon.**

13. A seed in an envelope at the store. **No—There is not enough moisture in the envelope.**

14. A seed in a moist paper towel in a sunny window. **Yes—All necessary conditions are present.**

Short answer:

15. Name the four organs of a plant. **Roots, Stem, Leaves, Flowers.**

CHALLENGE QUESTIONS

Mark each statement as either True or False.

16. _T_ Many plants can be used to make medicines.

17. _F_ All grass is alike.

18. _T_ Rye, wheat, and oats are all grasses.

19. _T_ Many trees have distinctive crowns.

20. _T_ A pine tree has a triangular growth habit.

21. _F_ Pruning will not affect a tree's growth habit.

22. _F_ External seed dormancy depends on temperature.

23. _T_ Some commercial growers use sulfuric acid to scarify seeds.

24. _T_ Stratification of seeds can occur in a refrigerator.

25. _F_ Seeds with double dormancy can experience scarification and stratification in either order and still germinate.

26 _F_ You will see cotyledons above ground if a plant experiences hypogeal germination.

27. _T_ Seeds absorb up to 200% more water when they germinate than they had before germination.

28. _F_ Seed dispersal is unimportant.

29. _T_ A deer can be a dispersing agent.

30. _F_ Only tiny seeds can float for water dispersal.

ROOTS & STEMS

LESSON 11

ROOTS

A GREAT FOUNDATION

SUPPLY LIST

Unpeeled carrot and any other roots you would like to examine   Magnifying glass
Jars with beans and corn from lessons 8 and 9
Supplies for Challenge: Radish seeds   Paper towels   Magnifying glass

BEGINNERS

- What are the four parts of plants? **Roots, stems, leaves, and flowers.**

- What are the two main jobs of the roots? **To hold the plant in place and to suck up water and nutrients for the plant.**

- What are the two different kinds of roots? **Taproots and fibrous roots.**

WHAT DID WE LEARN?

- What are the four organs of a plant? **Roots, stem, leaves, and flowers.**

- What are the jobs that the roots perform? **Anchoring the plant, absorbing water and nutrients, and storing extra food.**

- How can you tell what kind of root system a plant has? **You can examine the roots. Also, look at the seeds to see how many cotyledons they have. Generally, monocots have fibrous roots and dicots have taproots. We will see in future lessons that monocots and dicots usually have different types of leaves, too. This can also give us a clue to what type of root structure a plant has.**

TAKING IT FURTHER

- What kind of plants might you want to plant on a hillside? Why? **You might want to plant grass because the fibrous root system will spread out and hold the soil in place. Plants with taproots would not help stop the soil from washing away as well as plants with fibrous roots.**

- Why are the roots of plants like carrots and beets good to eat? **The leaves produce more energy than the plant can use at one time so the extra energy is converted and stored in the roots. We can eat those roots and get that energy for our bodies.**

LESSON 12 · SPECIAL ROOTS

NOT ALWAYS UNDERGROUND

SUPPLY LIST

Onion with roots   Flower bulbs (tulips, daffodils, etc.), if available

Supplies for Challenge: Poster board   Pictures of special roots   Drawing materials

BEGINNERS

- How do the roots on tulip bulbs look? **Like little hairs growing out from the bottom.**
- How do plants that do not grow in soil get water? **They have special roots that can take water out of the air.**
- What special kind of roots does a mangrove tree have? **Prop roots.**

WHAT DID WE LEARN?

- What are adventitious roots? **Roots that grow individually from the stem or bulb.**
- What are aerial roots? **Roots that grow in the air.**
- What are prop roots? **Roots that grow outward from the side of the stem, then downward to provide additional support.**

TAKING IT FURTHER

- Why do you think that some plants have specialized roots? **Not all environments are equally friendly to plant growth. God designed plants to grow in many areas that are hostile to most plants.**
- Why do some plants need prop roots? **Plants that grow in very soft or often wet soil may not be anchored well enough with a single root system. They need prop roots to give additional support.**

LESSON 13 · STEMS

CONNECTING IT ALL TOGETHER

SUPPLY LIST

Stalk of celery with leaves   Glass of water   Food color (red or blue)

2 clear plastic cups with potting soil   Jars with beans and corn from lessons 8 and 9

BEGINNERS

- What is the main job of the stem? **To transport water and food.**
- How are tree stems different from wildflower stems? **They are stiff and woody and covered with bark instead of being bendable.**
- How does the water and food move through the stem? **Through a series of special tubes.**

What did we learn?

- What are the main functions of a stem? **To support the plant, to carry nutrients and water between the roots and the leaves and flowers, to carry food from the leaves to the rest of the plant.**

- What do we call the stem of a tree? **A trunk or a branch.**

Taking it further

- If a branch is 3 feet above the ground on a certain day, how far up will the branch be 10 years later? **It will still be 3 feet above the ground. The tree will get taller but the growth is as the top of the tree and the ends of the branches, so the base of that particular branch remains in the same location.**

- What are some stems that are good to eat? **Celery is a good stem. Potatoes are special stems called tubers. Onions are special stems called bulbs. These are a few examples of stems that are good to eat.**

LESSON 14
STEM STRUCTURE
How they are put together

Supply list

Drawing materials   Plant with new growth, if available

Beginners

- Branches are what part of a plant? **The stem.**
- Name two things, besides branches, that grow from the stem. **Leaves and flowers.**

What did we learn?

- What are the major structures of a stem? **Shoot, terminal bud, lateral bud, node, and internode.**
- Where does new growth occur on a stem? **Most of the growth occurs at the terminal bud. New shoots, leaves, and flowers grow at lateral buds.**
- What gives the plant its size and shape? **The collection and arrangement of stems.**

Taking it further

- What will happen to a plant if its terminal buds are removed? **It will stop growing. It may form new terminal buds on new shoots, but the existing stems will be unable to grow.**
- How are stems different between trees and bushes? **Trees have one main stem or trunk with many smaller stems branching off. Bushes have many stems that all grow from the roots.**
- In your experience, do flower stems have the same structures, including terminal buds, nodes, etc., as bush and tree stems? **Most plants have the same structures you learned about here. Sometimes the stems are very short and it may be difficult to identify all of the structures, but most of flowering plants have the same stem structures.**

LESSON 15

STEM GROWTH

FURTHER UP AND FURTHER OUT

SUPPLY LIST (none)

BEGINNERS

- Why do tree trunks get bigger around every year? **New cells are formed inside the trunk and they push out on the bark.**

- When are new cells formed inside the trunk? **Mostly during the spring and summer.**

- What are growth rings? **Light and dark colored rings formed from new cells inside the trunk during growing and resting periods.**

WHAT DID WE LEARN?

- What are epidermis cells? **The cells on the outside of a young stem.**

- What is bark? **The epidermis cells that have been pushed outward, hardened, and died.**

- Name three types of cells inside a stem. **Xylem, phloem, and cambium cells.**

TAKING IT FURTHER

- Can we tell a tree's age from the rings inside the trunk? Why or why not? **Different cells are produced during different parts of the growing season, and few cells are produced during the winter so each year one set of colored bands is produced inside the trunk or stem of the tree. However, under certain conditions, multiple rings can form in a single year.**

- If you wanted to make a very strong wooden spoon, which part of the tree would you use? **The center, where the heartwood is.**

- Why don't herbaceous plants have bark? **They die at the end of each growing season so there is no time for bark to develop.**

CHALLENGE: VASCULAR TISSUE

- In lesson 13, you watched fluids moving up the stem of a stalk of celery. Do you remember how the xylem were arranged in the celery? They were arranged in a circular pattern. Would that indicate that celery is a monocot or a dicot? **Celery is a dicot.**

QUIZ 3

ROOTS & STEMS

LESSONS 11–15

Choose the best answer for each statement or question.

1. _B_ Which is not a function of the roots of a plant?
2. _A_ Which is not considered a plant organ?

3. _C_ A plant with this type of roots is most likely to live where it is dry.

4. _D_ A plant with this type of roots is most likely to live in a tree.

5. _A_ You would be most successful planting plants with these roots on a steep hill.

6. _B_ A plant with these kind of roots will be more successful in a very wet area.

7. _A_ Which is not a function of the stem?

8. _C_ What causes water to move upward in a plant?

9. _D_ What is a new stem called?

10. _C_ Where do leaves connect to the stem?

11. _B_ Which kind of cells are not found inside a stem?

12. _A_ Which cells carry water up a stem?

13. _C_ Which cells carry food down a stem?

14. _B_ Which cells protect a young stem?

15. _D_ Which cells protect mature stems?

CHALLENGE QUESTIONS

Choose the best answer for each question.

16. _B_ How does primary growth change the root?

17. _A_ Where does primary growth occur in a root?

18. _B_ How is osmosis different from diffusion?

19. _D_ What role does capillarity play in plants?

20. _C_ Which type of branching results in a wide low tree or shrub?

21. _A_ How are vascular bundles arranged in an herbaceous monocot stem?

22. _D_ What type of cells produce new xylem and phloem?

LEAVES

LESSON 16

PHOTOSYNTHESIS

MAKING FOOD FOR THE WORLD

SUPPLY LIST

3 mint plants or other fast-growing plants   Copy of "Photosynthesis Data Sheet"   Scissors
2 cardboard boxes that are big enough to cover a plant and allow for growth   Liquid measuring cup
Supplies for Challenge: Copy of "Photosynthesis Building Blocks" worksheet   Scissors   Tape

BEGINNERS

- What color is chlorophyll? **Green.**
- What is photosynthesis? **It is the process in which leaves turn water, carbon dioxide, and light into sugar and oxygen. It is the way plants make food.**
- What do plants produce that helps animals breathe? **Oxygen.**

WHAT DID WE LEARN?

- What are the "ingredients" needed for photosynthesis? **Water, carbon dioxide, chlorophyll, and sunlight.**
- What are the "products" of photosynthesis? **Food for the plant in the form of sugar and oxygen.**
- How did God specifically design plants to be a source of food? **Plants make their own food using the energy from the sun. They produce more than they need so the extra food energy is passed on to the animal or human that eats it.**
- How does carbon dioxide enter a leaf? **Through holes in the leaf called stomata.**

TAKING IT FURTHER

- On which day of creation did God create plants? **Day 3—Genesis 1:9–13.**
- On which day did He create the sun? **Day 4—Genesis 1:14–19.**
- Some people try to combine the Bible and evolution by saying that there were long periods of time between each day of creation. If this were true, what would have happened to all the green plants if there were a very long period of time between Day 3 and Day 4? **They all would have died.**
- In our experiment, we found that the plant that got less sunlight grew more slowly than the one that had full sunlight. Is this true for all plants? **No, many plants prefer the shade to full sun. God designed these plants to grow where other plants do not. Consider repeating this experiment with a shade-loving plant such as impatiens.**

LESSON 17

ARRANGEMENT OF LEAVES

MAXIMIZING SUNLIGHT

SUPPLY LIST

Drawing paper   Crayons, markers, or colored pencils
Supplies for Challenge: Aloe plant or other succulent, if available

BEGINNERS

- Explain different ways that leaves can grow from stems. **Answers will vary.**
- What is the main job of the leaves? **To perform photosynthesis/make food for the plant.**
- Why are leaves arranged in different ways? **To allow the sun to shine on them most of the day.**

WHAT DID WE LEARN?

- What are four common ways leaves can be arranged on a plant? **Opposite, alternate, whorled, and rosette.**
- Why do you think God created each of these different leaf arrangements? **Each of the leaf arrangements helps to ensure that one leaf does not block the light from reaching another leaf. The different arrangements are efficient for the size and shape of the leaves.**
- Why is it important for sunlight to reach each leaf? **Leaves need sunlight for photosynthesis. This is what keeps the plant alive and growing.**

TAKING IT FURTHER

- How does efficient leaf arrangement show God's provision or care for us? **Maximizing food production in plants provides more food for all animals, as well as for humans.**
- What other feature, besides leaf arrangement, aids leaves in obtaining maximum exposure to sunlight? **Leaves turn to follow the sun. Cells away from the sun become longer than those on the sunny side, allowing the leaf to turn toward the sun as it moves through the sky.**

LESSON 18

LEAVES—SHAPE & DESIGN

WHAT'S YOUR SHAPE?

SUPPLY LIST

1 or more large leaves freshly picked from a tree   Knife or scissors   Paper and colored pencils
Red food coloring   Bean and corn plants from lessons 8 and 9   Grass

BEGINNERS

- What shape of leaves do most plants have? **Broad leaves—wide and flat.**
- What shape are grass leaves? **Long, thin/narrow, and flat.**

- How do food and water move in leaves? **Through veins.**
- What do veins look like in grass? **Long, parallel lines.**
- What do veins look like in broad leaves? **One main vein with smaller veins branching off.**

OBSERVING LEAF SHAPES AND VEIN ARRANGEMENTS

- How are their shapes and vein arrangements different? **Corn—long, thin, parallel; Bean—broad, palmate.**
- Which plant has broad leaves? **Bean.**
- Which has long narrow leaves? **Corn.**
- Which plant is a monocot? **Corn.**
- Which plant is a dicot? **Bean.**

WHAT DID WE LEARN?

- What general shape of leaves do monocots and dicots have? **Monocots usually have long narrow leaves with parallel veins. Dicots usually have wider flat leaves with either pinnate or palmate veins. Evergreens usually have needles or scales for leaves.**

- How can we use leaves to help us identify plants? **Each plant has leaves with a unique shape and pattern. Some, such as the maple leaf, are very distinctive and easily recognizable. Others, grass for instance, can be more generic but can still be used to identify the species of plant.**

- How do nutrients and food get into and out of the leaves? **Xylem brings nutrients into the leaf, and then after the leaf has performed photosynthesis, the phloem transports the sugar from the leaves to the rest of the plant.**

TAKING IT FURTHER

- Describe how the arrangement of the veins is most efficient for each leaf shape. **Long, narrow leaves don't need veins that branch out, so parallel veins work well. Wider leaves need veins that reach out to the whole leaf, so palmate and pinnate veins are needed. For example, a maple leaf is nearly as wide as it is long, so a palmate arrangement of leaves is most efficient for transporting nutrients.**

LESSON 19

CHANGING COLORS

THE BEAUTY OF AUTUMN

SUPPLY LIST

If it is autumn, collect different colored leaves. If not, use several colors of construction paper
Scissors   Glue   Tag board/poster board   Newspaper   Heavy book
Supplies for Challenge: 2 or 3 different fresh leaves   Fingernail polish remover
Coffee filters   Quarter or other coin   Tape   Dish

BEGINNERS

- Why do trees lose their leaves in the fall? **To protect the trees from the winter cold.**
- How do trees know when it is time to lose their leaves? **There is less sunlight in the fall.**

- What causes the leaves to change colors in the fall? **The trees quit sending water up to the leaves, so the chlorophyll gets used up and other colors in the leaves become visible.**

What did we learn?

- How do trees know when to change color? **The reduced amount of sunlight available each day signals the tree to begin preparing for winter.**
- Why do trees drop their leaves? **As protection from the cold weather.**
- Why don't evergreen trees drop their leaves in the winter? **Their leaves are not easily damaged by the cold so the tree can survive the winter without losing them.**

Taking it further

- Do trees and bushes with leaves that are purple in the summer still have chlorophyll? **Yes, but in smaller amounts compared to the red pigment.**
- What factors, other than daylight, might affect when a tree's leaves start changing color? **Temperature can affect it to some extent. Also, the amount of water available has some effect, but even when the fall is unusually warm the leaves still begin to change about the same time each year because of the shorter period of daylight.**

LESSON 20
TREE IDENTIFICATION
HOW DO I KNOW WHAT TREE IT IS?

Final Project supply list

Tree field guide such as *Peterson's First Guide to Trees, Reader's Digest Field Guide of North America,* or *Trees of North America* by C. Frank Brockman

Zipper bags or other storage containers   Access to trees with leaves   Index cards

Colored pencils   Leaf press or heavy books and newspaper   Colored paper

Photo album with magnetic pages or 3-ring binder with plastic sheet protectors

Supplies for Challenge: Growth habit drawings from lesson 7

Beginners

- Why can you use leaves to identify a tree? **Every tree's leaves have a unique shape.**

What did we learn?

- What are some ways you can try to identify a plant? **By its leaves, flowers, fruit, bark, etc.**
- What are the biggest differences between deciduous and coniferous trees? **Deciduous trees have flowers, broad leaves and lose all their leaves each fall. Coniferous trees have cones and needles, and do not lose their needles each fall.**

Taking it further

- Why do we need to be able to identify trees and other plants? **To safely identify poisonous plants, to recognize ecosystems, for fun, to appreciate the diversity and wonder of God's creation and to make good landscaping choices.**

QUIZ 4

LEAVES

LESSONS 16–20

1. What is the purpose of the stomata in leaves? **To allow carbon dioxide in and oxygen out.**

2. For each leaf below, describe its vein arrangement (palmate or pinnate) and identify it if you can.
   A. Palmate—Maple   B. Pinnate—Oak   C. Pinnate—Ash   D. Pinnate—Holly

Short answer:

3. Which leaf above appears to be a compound leaf? **C.**

4. What kind of leaf arrangement does plant D have? **Alternate.**

5. How do leaves follow the sun? **The tips of the leaves detect the light and send out a chemical which makes the cells on the shady side get longer, thus turning the leaf toward the sun.**

6. What makes leaves green? **The chlorophyll in the chloroplasts.**

7. Identify the "ingredients" (beginning materials) and the "products" (ending materials) of photosynthesis. Ingredients: **Light, water, carbon dioxide, and chlorophyll;** Products: **Glucose/sugar and oxygen.**

CHALLENGE QUESTIONS

Short answer:

8. For each leaf above describe the leaf margin. **A. Lobed   B. Lobed   C. Toothed   D. Toothed**

9. Name two pigments that could be found in leaves. **Chlorophyll, carotene, xanthophyll, anthocyanin, etc.**

10. What is the chemical formula for photosynthesis? $6\,CO_2 + 6\,H_2O + light = C_6H_{12}O_6 + 6\,O_2$

11. What is the purpose of a bract? **To attract pollinators to the flowers.**

12. What is one purpose of a succulent leaf? **To store water, to perform photosynethesis.**

UNIT 5
FLOWERS & FRUITS

LESSON 21 FLOWERS

THE BEAUTY OF SIGHT AND SCENT

SUPPLY LIST

Copy of "Flower Pattern"   Green pipe cleaners   Flexible soda straws   Modeling clay
Colored construction paper   Scissors   Glue   Hole punch   Corn meal or yellow sand

BEGINNERS

- What are the four important parts of a plant? **Roots, stems, leaves, and flowers.**
- What are the four main parts of a flower? **Sepals, petals, stamens, and pistils.**

WHAT DID WE LEARN?

- What are the four parts of the flower and what is the purpose or job of each part? **The sepal protects the developing bud, the petals attract pollinators, the stamen produces pollen, and the pistil produces ovules that grow into seeds.**

TAKING IT FURTHER

- Why do you think God made so many different shapes and colors of flowers? **We enjoy the variety, and this shows us God's amazing creativity. Also, different animals are attracted by different colors and different scents. Hummingbirds are mainly attracted by red flowers, while other animals prefer different colors.**

LESSON 22 POLLINATION

THE BUZZING BEE'S JOB

SUPPLY LIST

Copy of the "Flip Book" worksheet   Crayons, markers, or colored pencils   Stapler
Bean and corn plants from lessons 8 and 9
Supplies for Challenge: Microscope and slide, if available   Pollen grains   Magnifying glass

BEGINNERS

- What is nectar? **A sweet liquid produced by the flower to attract insects.**

- What is pollination? **Moving pollen from one flower to another.**
- How does pollen get from one flower to another? **Bees and other insects move from flower to flower looking for nectar, and the pollen gets stuck to their bodies.**

WHAT DID WE LEARN?

- What animals can pollinate a flower? **Bees, wasps, moths, hummingbirds, bats, beetles, and even some small rodents can all be pollinators.**
- How can a flower be pollinated without an animal? **The wind or rain can move the pollen from the stamen to the pistil.**
- Does pollen have to come from another flower? **Not always, but it generally comes from another flower on another plant.**

TAKING IT FURTHER

- Why do you suppose God designed most plants to need cross-pollination? **The genetic information is stored in the pollen and ovules. If plants were always self-pollinated, much genetic information would be lost. The seeds formed from cross-pollination combine the hereditary traits of both parents, and the resulting offspring generally are more varied and often healthier than would be the case with self-pollination.**

LESSON 23 FLOWER DISSECTION
SEEING WHAT'S INSIDE

SUPPLY LIST

A fresh flower with easily-visible reproductive parts (A lily or an alstroemeria is a good example.)
Sharp knife or razor blade
Supplies for Challenge: Composite flower such as daisy, sunflower, or zinnia

BEGINNERS

- What is the job of the sepals? **To protect the developing flower.**
- What is the job of the petals? **To attract insects.**
- What is the job of the stamens? **To produce pollen.**
- What is the job of the pistil? **To receive pollen and make seeds.**

WHAT DID WE LEARN?

- How many ovules did you find? **Answers will vary.**
- What did they look like? **Answers will vary.**

TAKING IT FURTHER

- Why are the ovules in the flower green or white when most seeds are brown or black? **They are not fertilized and are not mature.**
- If you planted the ovules, would they grow into a plant? **No, they are not pollinated and are not mature or ready to grow into a plant.**

LESSON 24

FRUITS

IS IT RIPE YET?

SUPPLY LIST

Apple, strawberry, pineapple (all fresh and whole, if possible)   Knife
Supplies for Challenge: Copy of "Fruit Classification" worksheet

BEGINNERS

- Name three fruits. **Answers will vary.**
- What is the job of the fruit? **To help the seeds get to a new location.**

WHAT DID WE LEARN?

- What is the main purpose of fruit? **To make sure the seeds are dispersed.**
- What are the three main groups of fruit? **Simple, aggregate, and multiple.**
- Describe how each type of fruit forms. **Simple fruit forms one fruit from one flower with one pistil. Aggregate fruit forms one fruit from one flower with several pistils. Multiple fruit forms one piece from several flowers with each fruit fusing together into a whole.**

TAKING IT FURTHER

- What is the fruit of a wheat plant? **The kernel of wheat that we make into flour.**
- Which category of fruit is most common? **Simple.**
- Why do biologists consider a green pepper to be a fruit? **They are the mature ovary of the plant. Any reproductive structure is a fruit. Other examples include beans, peas, tomatoes, and broccoli.**

CHALLENGE: FRUIT CLASSIFICATION WORKSHEET

1. _D_ Acorn
2. _E_ Pea
3. _C_ Pear
4. _A_ Avacado
5. _A_ Mango
6. _E_ Peanut
7. _E_ Lima bean
8. _F_ Wheat
9. _A_ Nectarine
10. _B_ Green pepper
11. _D_ Cashew
12. _C_ Crabapple
13. _B_ Grapefruit
14. _F_ Corn
15. _F_ Rice

LESSON 25

ANNUALS, BIENNIALS, & PERENNIALS

HOW LONG DO THEY GROW?

SUPPLY LIST

Copy of "Plant Word Search"

BEGINNERS

- How long does it take for an annual plant to complete its lifecycle? **One year or one growing season.**

- What kinds of plants grow year after year? **Perennials, or trees and shrubs.**

PLANT WORD SEARCH

```
P H O T O S Y N T H E S I S S
H R S E E D S K Q E R T A L P
O G T E E S A A N N U A L V A
S T E R B T M C H R Y S E E S
T B M P H O R O S Y N I A C H
C H L O R O P H Y L L M V O P
Z O P O N P U P U W R F E T E
L H D R M A B I P R U L S Y R
V I P O L L I N A T I O N L E
O R A O C M E N T L A W Q E N
P I N T E A N A I P T E C D N
A B C S E T N T B A I R O O I
L E V B W E I E C C A S T N A
M A N V C P A R E S N I A L L
A F R U I T L D E M F R I U R
```

WHAT DID WE LEARN?

- What is an annual plant? **One that completes its lifecycle in one growing season.**

- What is a biennial plant? **One that completes its lifecycle in two growing seasons.**

- What is a perennial plant? **One that grows year after year, producing flowers and seeds each season.**

TAKING IT FURTHER

- Why don't we often see the flowers of biennial plants? **We usually harvest them the first season.**

- Why don't people grow new plants from the seeds produced by the annuals each year? **Often the conditions are not right for germination of the seeds produced the previous year. Also, many people clear their gardens of the dead flowers before the flowers have a chance to deposit their seeds. Finally, nurseries and greenhouses can begin growing plants inside much earlier than the plants would begin growing in the garden. This allows plants to be mature enough to be blooming by the beginning of spring. Plants that come up naturally in your garden would not bloom until much later in the summer, and people want to have blossoms in their gardens all spring and summer.**

QUIZ 5
FLOWERS & FRUITS

LESSONS 21–25

1. Match the labels to the parts of the flower below: a. **Pollen**, b. **Ovule**, c. **Ovary**, d. **Pistil**, e. **Petals**, f. **Stamen**, g. **Sepal**

Short answer:

2. Which part of the flower is considered the male part? **Stamen.**

3. Which part of the flower is considered the female part? **Pistil.**

4. How is a simple fruit different from an aggregate fruit? **Simple fruit forms one fruit from one flower with one pistil. Aggregate fruit forms one fruit from one flower with multiple pistils.**

5. Describe the process of pollination: **Pollen is removed from a stamen, usually by a pollinator like a bee or other insect. It is then deposited on the pistil of another flower. A pollen tube grows down into the ovary until it reaches the ovule. Fertilization takes place and the ovule becomes a seed.**

6. List two ways that pollen can be transferred from one flower to another. **Possible answers include: By an animal or other pollinator, by wind, rain, or by a person.**

7. How long does it take for a biennial to complete its lifecycle? **2 years or 2 growing seasons.**

CHALLENGE QUESTIONS

Mark each statement as either True or False.

8. _F_ Composite flowers have only one flower per stalk.

9. _T_ Ray flowers often look like petals.

10. _T_ Disk flowers produce hundreds of seeds.

11. _F_ Flowers do not need to attract pollinators.

12. _T_ Some nectar guides can normally only be seen by insects.

13. _T_ Some flowers smell bad to attract flies as pollinators.

14. _T_ Succulent fruits are simple fruits.

15. _T_ An olive is considered a fruit.

16. _F_ Peanuts are nuts.

17. _F_ Apples are berries from a biological definition.

18. _T_ Oranges are berries from a biological definition.

19. _F_ Ephemeral plants live for many years.

20. _T_ Ephemeral plants often live in the desert.

UNIT 6
UNUSUAL PLANTS

LESSON 26
MEAT-EATING PLANTS

WILL IT EAT ME?

SUPPLY LIST

Small box   Stick or pole
Supplies for Challenge: Drawing materials

BEGINNERS

- What is the most famous meat-eating plant? **Venus flytrap.**
- How does a butterwort trap insects? **It has sticky leaves.**
- Why do some plants eat insects? **They do not get enough nutrients from the soil.**

WHAT DID WE LEARN?

- What is a carnivorous plant? **One that eats animals, usually insects.**
- Why do some plants need to be carnivorous? **Some plants grow in areas where there are not enough nutrients in the soil. They trap insects to get the necessary nutrients to survive.**
- How does a carnivorous plant eat an insect? **It traps the insect then secretes an acid that breaks down the animal's body so the nutrients can be absorbed.**

TAKING IT FURTHER

- Where are you likely to find carnivorous plants? **In wet, marshy areas.**
- How might a Venus flytrap tell the difference between an insect on its leaf and a raindrop? **The flytrap is designed with trigger bristles that have to be moved in order for the leaf to close. Two or more must be moved within a short period of time for the leaf to close. A raindrop might touch one but would probably not trigger two or more, but a moving insect would.**

LESSON 27
PARASITES & PASSENGERS

LIVING OFF OF EACH OTHER

SUPPLY LIST

Drinking straw   Coffee stirrer (a narrow straw)   Field guide to plants   Knife or scissors   Sink

Supplies for Challenge: Research materials—depends on your chosen topic

BEGINNERS

- What is a parasitic plant? **One that steals water and nutrients from another plant.**
- What is a passenger plant? **One that grows on the side of another plant without hurting it.**
- How does a passenger plant get water? **With special roots that absorb water from the air.**

WHAT DID WE LEARN?

- What is a parasitic plant? **One that gets its nourishment from another plant instead of making its own food.**
- What is a passenger plant? **One that lives on the outside of another plant without harming it.**
- How do passenger plants obtain water and minerals? **They absorb them from the air and from the surface of the host.**

TAKING IT FURTHER

- Where is the most likely place to find passenger plants? **On trees—often in the rain forest, but also in other areas.**
- Do passenger plants perform photosynthesis? **Yes, they still make their own food.**
- Do parasitic plants perform photosynthesis? **Some do but most do not. They get their food from the host plant.**

LESSON 28

TROPISMS

HOW PLANTS RESPOND

SUPPLY LIST

A houseplant with several leaves

Supplies for Challenge: Copy of "Tropisms" worksheet

BEGINNERS

- Will a plant grow upside down if the seed is planted upside down? **No, it will sense up and down and grow correctly.**
- What are some things that plants have the ability to sense? **Up and down, water and light.**
- What will a plant do if something blocks the light? **It will try to grow around it.**

WHAT DID WE LEARN?

- What is geotropism? **The ability of plant roots to always grow down and stems to grow up, a response to gravity.**
- What is hydrotropism? **The ability of plant roots to grow toward water, a response to water.**
- What is phototropism or heliotropism? **The ability of plant leaves to turn toward the sun or a light source, a response to light.**

Taking it further

- Why are tropisms sometimes called "survival techniques"? **They allow the plant to survive even if conditions change. They give the plant a better chance for survival even when water is scarce or something blocks the sun.**

- Will a seed germinate if it is planted 5 feet (1.5 m) from the water? **No, seeds need water to soften the seed coat and germinate. Tropisms only help the plant after germination.**

- Where are some places you would not want to plant water-seeking plants such as willows? **Near a swimming pool, septic tank, or water line.**

LESSON 29
SURVIVAL TECHNIQUES
SURVIVING IN HARSH CLIMATES

Supply list

Cactus plant   Magnifying glass

Supplies for Challenge: Copy of "Designed for Survival" worksheet

Beginners

- Name three ways that God has designed plants to be able to survive harsh conditions. **Cactus can store water in its stem; cactus has needles that do not lose water; mountain plants grow close to the ground to avoid wind; mountain plants grow in groups to keep each other warm.**

What did we learn?

- How do some plants survive in hot, dry climates? **They quickly absorb the available water and store it in their expandable stems. They have needles instead of regular leaves so water does not evaporate quickly.**

- How do some plants survive in cold windy climates? **They have small leaves and short stems to withstand the wind. They grow low to the ground and in groups. They can reproduce very quickly.**

Taking it further

- Why do alpine plants need protection from the sun? **High in the mountains there is less atmosphere, so the sun's rays are more intense.**

Challenge: Designed for Survival worksheet

- List six things that plants need to survive. **Light, warmth, water, carbon dioxide, oxygen, trace elements including nitrogen and phosphorus, and a place to grow.**

- List six things that can harm plants. **Wind, lack of water, hail, insects, diseases, over crowding, parasites, lack of light, and over watering.**

- List 12 ways that plants have been designed to survive. **Broad leaf trees lose their leaves in winter, seeds do not germinate until conditions are favorable for growth (seed dormancy), plants have seed distribution techniques, seeds store energy for the growing shoot, special roots including prop roots, aerial roots, pneumatophore roots, photosynthesis, leaf arrangement, leaf shape, flower shape, meat-eating plants' designs, special stems including tendrils, thorns, stolons and runners, parasitic designs, passenger plants can use other plants to help them survive without harming them, all the various tropisms, ability to** store water, needle-like leaves on cacti, design of alpine plants, bracts attract pollinators, scent, color, and pollen guides help attract pollinators .

- List four ways that people help plants to survive. **Water your grass, flowers, or other plants, add fertilizer to the soil, plant certain plants in a favorable location such as in the shade or sun depending on the plant, provide a trellis for climbing plants, pulling weeds to prevent competition, spray insecticide or fungicide, remove parasites, prune.**

REPRODUCTION WITHOUT SEEDS

THERE ARE OTHER WAYS

SUPPLY LIST

Potato   Jar   Potting soil   Water

Supplies for Challenge: Research materials on genetic modification

BEGINNERS

- Where do most new plants come from? **Seeds.**
- How do we get new strawberry plants? **Older plants send out special stems call runners that grow new plants.**
- How do we get new tulips? **The bulbs underground grow more bulbs.**

WHAT DID WE LEARN?

- What are some ways that plants can reproduce without growing from seeds? **Some plants reproduce by sending out runners, producing bulbs, or growing new plants from parts cut from the original plant.**

TAKING IT FURTHER

- Why can a potato grow from a piece of potato instead of from a seed? **All of the growth information is located in the eyes of the potato, so a new plant can grow from this area.**
- Will the new plant be just like the original plant? **Genetically, the new plant will be identical to the original plant. This is not the case with plants grown from seeds. Seeds contain genetic information from both the plant producing the pollen and the plant producing the ovule, but plants grown vegetatively only get genes from the original plant.**

FERNS

SEEDLESS PLANTS

SUPPLY LIST

Paper   Paint and paint brushes   Glue   Corn meal or yellow sand
Fresh fern frond, if available

BEGINNERS

- How do plants without flowers grow new plants? **They produce spores.**
- What is one plant that produces spores? **Ferns.**
- Where do ferns grow? **In areas with lots of rain.**

WHAT DID WE LEARN?

- How are ferns like other plants? **They have chlorophyll, stems, leaves, and roots.**
- What are fern leaves called? **Fronds.**
- How are ferns different from other plants? **They do not have flowers or seeds.**
- How do they reproduce? **They make spores on the back of their fronds that produce an egg and sperm that combine and grow into a tiny new plant.**

TAKING IT FURTHER

- Why can't ferns reproduce with seeds? **They do not produce flowers, so they cannot make seeds.**

LESSON 32

MOSSES

DO YOU REALLY FIND MOSS ON THE NORTH SIDE OF TREES?

SUPPLY LIST

Paper and glue   Colored pencils or crayons   Dried moss from a craft shop   Magnifying glass
Supplies for Challenge: Peat mos   Dirt or soil from your yard   Paper cups

BEGINNERS

- How big are moss plants? **They are very small.**
- What plant parts do mosses have? **Leaves and stems and something like roots.**
- How do moss plants reproduce? **With spores.**
- Where do moss plants grow? **Anywhere that is wet.**

WHAT DID WE LEARN?

- How do mosses differ from seed-bearing plants? **They have no flowers, seeds, or true roots.**
- How do mosses differ from ferns? **They are smaller, have no true roots, and produce their spores on stalks instead of on their leaves.**
- How do mosses produce food? **They have chlorophyll and perform photosynthesis just like other plants do.**

TAKING IT FURTHER

- Are you likely to find moss in a desert? Why/why not? **No. There is not enough moisture in the desert for moss to grow well.**

LESSON 33

ALGAE

ARE ALL GREEN THINGS PLANTS?

SUPPLY LIST

Paper   Colored pencils   Glue   Construction paper   Scissors
Supplies for Challenge: Sample of pond water   Microscope and slide

BEGINNERS

- How are algae similar to plants? **They have chlorophyll and perform photosynthesis, but they have no leaves, stems, or roots.**
- Why are algae important? **They produce food for the sea creatures and most of the oxygen on the planet.**

WHAT DID WE LEARN?

- Why are algae such important organisms? **They produce large amounts of oxygen and they are the major source of food in many aquatic food chains.**
- What gives algae its green color? **Chlorophyll.**

TAKING IT FURTHER

- Why are some algae yellow, brown, blue, or red? **All algae have chlorophyll, but like many other plants, some have other pigments as well, which often cover up the green of the chlorophyll.**

LESSON 34

FUNGI

ARE THESE REALLY PLANTS?

SUPPLY LIST

6 slices of bread (homemade works best)   3 plastic sandwich bags or plastic zipper bags
Copy of "Mold Data Sheet"
Supplies for Challenge: Fresh mushroom   Index card   Aerosol hairspray

BEGINNERS

- Are mushrooms plants? **No, they are fungi.**
- Name three kinds of fungi. **Mushrooms, mold, and yeast.**
- What is one way that fungi can be bad? **Some are poisonous, some cause disease, and some spoil food.**
- What is one way that fungi can be good? **Some are good to eat, yeast makes bread fluffy, some give cheese its special flavor, some get rid of dead plants and animals.**

WHAT DID WE LEARN?

- Why are fungi not considered plants and given their own kingdom? **Fungi do not have chlorophyll and do not have roots, stems, and leaves as plants do.**

- What are some good uses for fungi? **Fungi are used for food, to make bread rise, to make medicines, to give cheese its flavor, and to help in the recycling of dead plants and animals.**

TAKING IT FURTHER

- What other conditions might affect mold growth other than those tested here? **Light/dark or the presence of chemicals like the preservatives found in foods.**

- How can you keep your bread from becoming moldy? **Keep it in a cool dry place.**

QUIZ 6

UNUSUAL PLANTS

LESSONS 26–34

Match the term with its definition.

1. _C_ Plants that eat insects
2. _B_ Plants that "steal" nutrients from other plants
3. _D_ Plants that grow on other plants without harming them
4. _A_ Response of plants to gravity, roots go down, stems go up
5. _E_ Tendency for roots to grow toward water
6. _G_ Ability of leaves to turn toward sunlight
7. _F_ Plant designed to store available water in dry conditions
8. _G_ Plant reproduction using a part of the plant (not seeds)
9. _H_ Runners from a strawberry plant
10. _I_ Special stems that grow underground for reproduction

Short answer:

11. What plant organ is missing in ferns? **Flowers.**
12. How do both mosses and ferns reproduce? **Spores.**
13. What two plant organs are missing in mosses? **Flowers and true roots.**
14. What substance do algae have in common with plants? **Chlorophyll.**
15. What is the name of the group that contains yeast and mushrooms? **Fungi.**

CHALLENGE QUESTIONS

Short answer: (Accept reasonable answers for all challenge questions.)

16. Give an example of positive tropism. **Hydrotropism, phototropism, and geotropism for roots.**

17. Give an example of negative tropism. **Geotropism for stems, thermotropism for curling leaves, and thigmotropism for roots.**

18. Explain how a cobra lily traps insects. **It attracts insects to its pitcher with nectar, then as the insect tries to find the exit it hits the top of the pitcher and falls inside.**

19. Explain how succulents are designed to survive dry periods. **They can store large amounts of water in their stems and/or leaves.**

20. Why is grafting a form of cloning? **The resulting plant has identical DNA to the parent plant.**

21. Where would you likely find the stems of a fern plant? **Underground.**

22. Give one reason why peat moss is important. **It holds water and increases the moisture of the soil, it can be used for fuel, and mosses create new soil and return nutrients to the soil .**

23. Give one commercial use of algae. **Scrubbing agent, makes many items creamy, used as a food, and thickens ice cream and other products.**

FINAL EXAM WORLD OF PLANTS

LESSONS 1–34

Define each of the following terms.

1. Geotropism: **Response of plant to gravity; causes roots to grow down and stems to grow up.**

2. Hydrotropism: **Response of plants to water; causes roots to grow toward a source of water.**

3. Phototropism: **Response of plants to light; leaves turn toward the sun.**

4. Photosynthesis: **Process by which sunlight, chlorophyll, carbon dioxide, and water are turned into sugar and oxygen.**

5. Pollination: **Process by which pollen is transferred from one flower to another to cause the fertilization of the ovule, thereby creating seeds.**

6. Chlorophyll: **The green substance in plant cells that is used to perform photosynthesis.**

7. Ovule: **The egg or unfertilized seed found in the ovary of the flower.**

8. Pistil: **The female part of the flower that produces ovules.**

9. Stamen: **The male part of the flower that produces pollen.**

10. Xylem and phloem: **The tubes that carry nutrients, food, and water throughout the plant.**

Choose the best answer for each question.

11. _C_ Which of the following is not an organ of flowering plants?

12. _D_ Which kind or kinds of creatures get nourishment from grasses?

13. _A_ Which is not a common use for the wood of a tree?

14. _B_ Which tree is a deciduous tree?

15. _C_ Which organ is primarily used to absorb minerals from the ground?

Fill in the blank with the correct term.

16. Root growth primarily occurs at the **_root tip or root cap_**.

17. The two types of root systems are **_fibrous_** and **_taproot_**.

18. The shape of most monocot plants' leaves is **_long and thin like grass_**.

19. Ferns reproduce by **_spores_** on their leaves.

20. Algae are similar to plants because they contain **_chlorophyll_**.

Mark each statement as either True or False.

21. _F_ Plants with red leaves have no chlorophyll.

22. _T_ Trees can be identified by their leaves.

23. _F_ Coniferous trees do not have leaves.

24. _F_ The scent of a flower has no purpose.

25. _T_ Ferns are not flowering plants.

26. _T_ Algae is an important organism.

27. _T_ Sepals might be confused with leaves.

28. _T_ Pollination must take place for seeds to form.

29. _T_ Mosses reproduce by spores and not seeds.

30. _T_ Photosynthesis cannot take place without chlorophyll.

CHALLENGE QUESTIONS

Match the term with its definition.

30. _E_ Law of biogenesis

31. _C_ Meiosis

32. _A_ Spontaneous generation

33. _D_ Scarification

34. _F_ Stratification

35. _B_ Seed dormancy

36. _K_ Primary growth

37. _J_ Secondary growth

38. _G_ Osmosis

39. _H_ Toothed leaf margin

40. _I_ Lobed leaf margin

Mark each statement as either True or False.

41. _F_ Ephemeral plants grow slowly.

42 _T_ Composite flowers are really hundreds of flowers grouped together.

43. _F_ Legumes have hard outer shells.

44. _T_ Pomes have papery inner cores.

45. _T_ Chemotropism aids in pollination.

46. _T_ Filament algae is very common.

47. _F_ Most fruit trees are grown from seeds.

48. _F_ Tendrils have negative tropism.

49. _T_ Rootstock is important for grafting.

50. _F_ There are very few commercial uses for algae.

CONCLUSION

APPRECIATING THE WORLD OF PLANTS

SUPPLY LIST

Tag board   Dried leaves, flowers, grass   Glue   Seeds, seed pods, other parts of plants

Resource Guide

Many of the following titles are available from Answers in Genesis (www.AnswersBookstore.com).

Suggested Books

Trees of North America by C. Frank Brockman—Great guide for tree identification

Reader's Digest North American Wildlife—A great resource to have for any field trip!

The Science of Plants by Jonathan Bocknek—Overview of plants, activities, puzzles for all ages

Plants by Janice VanCleave—Lots of fun activities; science fair ideas

Biology for Every Kid by Janice VanCleave—More fun activities

The Nature and Science of Seeds by Jane Burton and Kim Taylor—Great pictures, overview of less common seeds

Photosynthesis by Alvin Silverstein—In-depth look at the process for upper elementary

How Leaves Change by Sylvia A. Johnson—Great pictures of fall colors, more in-depth information for upper elementary students

God Created the Plants and Trees of the World by Earl & Bonita Snellenberger—Coloring and sticker book for children that teaches about plants and trees from a biblical perspective

Suggested Videos

Newton's Workshop by Moody Institute—Excellent Christian science series; several titles to choose from

Field Trip Ideas

- Creation Museum in Petersburg, KY
- Botanic gardens
- Tour a local nursery, greenhouse, or florist
- Arboretum
- Take a nature walk

Creation Science Resources

Answers Book for Kids Four volumes by Ken Ham with Cindy Malott—Answers children's frequently asked questions

The New Answers Books 1 & 2 by Ken Ham and others—Answers frequently asked questions

Dinosaurs by Design by Duane T. Gish—All about dinosaurs and where they fit into creation

The Amazing Story of Creation by Duane T. Gish—Scientific evidence for the creation story

Creation Science by Felice Gerwitz and Jill Whitlock—Unit study focusing on creation

Creation: Facts of Life by Gary Parker—In-depth comparison of the evidence for creation and evolution

Dinosaurs for Kids by Ken Ham—Learn the true history of dinosaurs

MASTER SUPPLY LIST

The following table lists all the supplies used for *God's Design for Life: World of Plants* activities. You will need to look up the individual lessons in the student book to obtain the specific details for the individual activities (such as quantity, color, etc.). The letter *c* denotes that the lesson number refers to the challenge activity. Common supplies such as colored pencils, construction paper, markers, scissors, tape, etc., are not listed.

Supplies needed (see lessons for details)	Lesson
Aloe plant	17c
Bread slices (homemade or with no preservatives)	34
Cactus plant	29
Cardboard boxes or shoe boxes	4, 16, 26
Coffee filters	19c
Coffee stirrer	27
Corn meal or yellow sand	21, 31
Craft sticks	6c
Dried moss (from craft store)	32
Encyclopedia (plant and animal)	3
Fern frond	31
Field guide (flowers)	5
Field guide (plants)	27
Field guide (trees)	20
Fingernail polish remover	19c
Flower (composite, such as daisy, sunflower, or zinnia—fresh)	23c
Flower (such as lily—fresh)	23
Flower bulbs (tulips, daffodils, etc.—optional)	12
Food coloring	13, 18
Fruits, nuts, and vegetables	4c, 12, 10, 11, 13, 24, 30, 34c
Gelatin mix (yellow)	4
Grass plant	6
Grapes (red and green)	4
Hairspray (aerosol)	34c

Supplies needed (see lessons for details)	Lesson
Index cards	7, 17, 20, 34c
Jars (1 must have a lid)	8, 9, 30
Knife or scalpel (very sharp)	9, 18, 23, 24
Leaves (fresh)	19c, 20
Magnifying glass	6, 9, 11, 20, 22c, 29, 32
Microscope and slides	4c, 22c, 33c
Modeling clay	21
Peat moss	32c
Photo album with magnetic pages or 3-ring binder	20
Pinecones (scales tightly shut)	10
Pipe cleaners	21
Plants (fast-growing; e.g., mint plants)	16, 28
Plastic cups (clear)	13, 32c
Plastic zipper bags	4, 20, 34
Pollen	22c
Pond water	33c
Poster board/tag board	2, 12c, 19, 35
Potting soil	6c, 13, 30
Seeds (bean, corn, grass, coconut, radish)	6c, 8, 9, 10c, 11c
Steel wool	8
Straws (flexible)	21, 27

WORKS CITED

Adams, A. B. *Eternal Quest: The Story of the Great Naturalists*. New York: G.P. Putnam's Sons, 1969.

Bocknek, Jonathan. *The Science of Plants*. Milwaukee: Gareth Stevens Publishing, 1999.

Brockman, C. Frank. *Trees of North America*. New York: Golden Press, 1968.

Burnie, David. *Eyewitness Books Plants*. New York: Alfred A. Knopf, 1989.

Burton, Jane, and Kim Taylor. *The Nature & Science of Seeds*. Milwaukee: Gareth Stevens Publishing, 1999.

"Cobra Lily." http://www.bugbitingplants.com/cobra_lily.php.

Challand, Helwn J. *Plants Without Seeds*. Chicago: Children's Press, 1986.

"A Concise History of the Rose—The King of Flowers." http://www.indiainternational.com.

Croll, Carolyn. *Redoute The Man Who Painted Flowers*. New York: G. P. Putnam's Sons, 1996.

'Espinasse, M. *Robert Hooke*. Berkeley: University of California Press, 1962.

Evans, Erv and Blazich, Frank. *Overcoming Seed Dormancy: Trees and Shrubs*. http://www.ces.ncsu.edu/depts/hort/hil/hil-8704.html.

"The Faces of Science: African Americans in the Sciences." http://webfiles.uci.edu/mcbrown/display/faces.html.

Giannotti, Heike. *The History of Rose Culture in France*. University of Illinois, 2001.

Giesecke, Ernestine. *Outside My Window Flowers*. Des Plaines: Heinemann Library, 1999.

Gish, Duane T. *The Amazing Story of Creation from Science and the Bible*. El Cajon: Institute for Creation Research, 1990.

Gish, Duane T. *Dinosaurs by Design*. Colorado Springs: Creation Life Publishers, 1992.

Greenaway, Theresa. *Mosses & Liverworts*. Austin: Steck-Vaughn Co., 1992.

Greenaway, Theresa. *The Plant Kingdom*. Austin: Steck-Vaughn Co., 2000.

Ham, Ken et al. *The Answers Book*. El Cajon: Master Books, 1992.

Jebens, Brandon. *The Biogeography of Sequoia Sempervirens*. San Fancisco: San Francisco State University, 1999.

Johnson, Sylvia A. *How Leaves Change*. Minneapolis: Lerner Publications Co., 1986.

Karaler, Lucy. *Green Magic Algae Rediscovered*. New York: Thomas Y. Crowell, 1983.

Koerner, L. *Linnaeus: Nature and Nation*. Cambridge: Harvard University Press, 1999.

Lindroth, S. *The Two Faces of Linnaeus*. Berkeley: University of California Press, 1983.

Moore, J. A. *Science as a Way of Knowing*. Cambridge: Harvard University Press, 1993.

Morris, John D., Ph.D. *The Young Earth*. Colorado Springs: Master Books, 1994.

Parker, Gregory, and others. *Biology: God's Living Creation*. 2nd ed. Pensacola: A Beka Book, 1986.

Omahen, Sharen, *Seed Storage Facility is Modern Day Noah's Ark*. http://www.uga.edu/discover/educators/readings/read187.pdf, April, 2003.

"Pierre-Joseph Redoute." http://www.globalgallery.com/bios/redoute.

Reader's Digest. *North American Wildlife*. Pleasantville: Reader's Digest Association, 1982.

Rogers, Kirsteen et al. *Usborne Science Encyclopdia*. London: Usborne Publishing, 2002.

Rudwick, M. J. S. *The Meaning of Fossils*. Chicago: University of Chicago Press, 1985.

Schiebinger, L. "The Loves of Plants." *Scientific American*. February 1996: 110-115.

Selsam, Millicent E. *Mushrooms*. New York: William Morrow & Co., 1986.

Silverstein, Alvin, and others. *Photosynthesis*. Brookfield: Twenty-first Century Books, 1998.

Steele, DeWitt. *Investigating God's World*. Pensacola: A Beka Book Publications, 1986.

"This Person in Black History 4: George Washington Carver." http://sps.k12.mo.us/historyday/feb/carver.htm.

"The Truly Amazing Redwood Tree." http://www.treesofmystery.net/sequoia.htm.

VanCleave, Janice. *Biology for Every Kid*. New York: John Wiley & Sons, Inc., 1990.

VanCleave, Janice. *Plants*. New York: John Wiley & Sons, Inc., 1997.

Wexler, Jerome. *From Spore to Spore*. New York: Dodd, Mead & Co., 1985.